# CONTENTS

**CANOPY** BOOKS

Written by Kris Hirschmann
Designed by Scott Westgard
© 2016 by Canopy Books, LLC
1001 2nd Ave #280, New Hyde Park, NY 11040

Made and Printed in
Foshan, China

Tracking # 9781946426055-1
ISBN: 978-1-946426-05-5
Ages 7 & Up

D1450875

# WHAT IS SLIME?

It's ooey, gooey, icky, and sticky. It slithers and slips; it pours and drips; it oozes and amuses. What is this amazing material? It's SLIME, of course!

Okay, so that question wasn't too tough. Just about anyone can recognize slime. But there's more to slime than meets the eye. Slime is actually a fascinating substance with lots of interesting scientific properties. That's where this kit comes in. You want to play with slime and talk about science? Guess what: You're in luck, because you can — and will — do both of these things at the very same time. Get ready to get SLIMED!

## Poly-What?

So let's get some of the science out of the way right now. What exactly is slime, anyway? In scientific terms, it's something called a **polymer**.

Polymers are substances with long, stringy **molecules**. The molecules are all tangled up together, like strands of spaghetti on a plate. Can these molecules move? Well, sometimes. It all depends on how bad the tangling is. Materials with *really* mixed-up molecules, such as hard plastics, can't budge at all. Materials with looser molecules, such as slime, can flow a little bit. They ooze from one place to another as the molecules slooooowly slide around.

And *ta-da*! There you have it. Slime. NOW you're seeing things clearly!

# IN THIS KIT

This kit comes with items you can use to slime things up. The instructions in this book tell you how and when to use these items. **For best results, always read the instructions carefully!**

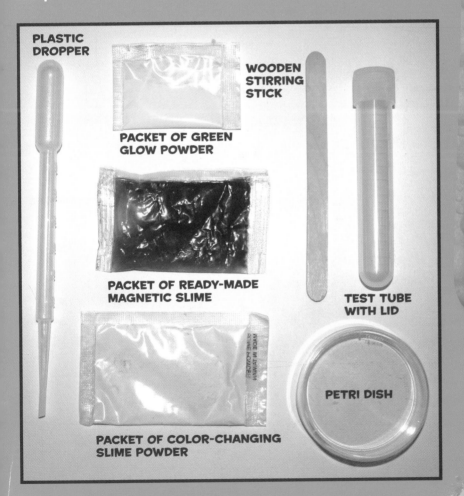

PLASTIC DROPPER

PACKET OF GREEN GLOW POWDER

WOODEN STIRRING STICK

PACKET OF READY-MADE MAGNETIC SLIME

TEST TUBE WITH LID

PACKET OF COLOR-CHANGING SLIME POWDER

PETRI DISH

# WHERE'S THE SLIME?

You know *what* slime is. But do you know *where* it hides? Believe it or not, you can find polymers almost anywhere!

## Toys and Entertainment

The toy store is one guaranteed source for slime. Just-for-fun slime comes in little cans, plastic packets, science kits (like this one), and countless other formats. It's made in practically any color, texture, and smell you can imagine. You can find slime on TV, too. One TV station is known for dumping sticky green goop onto game-show guests.

## Industrial Slime

Slime isn't just for fun. It's practical, too! Slimy polymers make eye drops slippery. Doctors use them to help stop bleeding during surgical procedures. But that's not all. Polymers also help give stretchy fabrics their "stretch." They even make concrete stronger! In daily life, there's no telling where slime might appear next.

## Up Your Nose

Okay, this example is a little gross. But there's no escaping the disgusting truth: Your body is FULL of slime. Mucus is one example. Whether it's oozing from your left nostril, bubbling through your digestive system, or lubricating your liver, this stuff is icky but important for every person's good health.

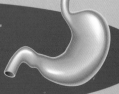

### SLIME SAVES YOUR LIFE

Your inner stomach is coated with a layer of slime. If it wasn't, it would digest itself. Really!

## In Nature

There's so much slime in nature, it's almost hard to know where to start. There are slimy animals (snails, slugs, eels, and so on). There are slime molds, which are oh-so-much slimier and grosser than regular hairy molds. There are slimy foods, like eggs. There are even slimy bacteria that form huge, sticky colonies of goo. All in all, it's safe to say that Mother Nature LOVES her slime!

## BE A SLIME SCIENTIST

Handling slime is a science. Here are a few tips that will guarantee your gooey success.

- **SET UP A LAB.** Most of the experiments in this book are messy. Lay down newspaper before you begin. This will keep your lab area clean.

- **NO EATING ALLOWED!** Do not put any slime near your mouth. Wash your hands with soap and water after you touch any slime.

- **DON'T PUT SLIME DOWN THE DRAIN.** If you need to get rid of your slime, put it into a plastic bag. Seal the bag, then throw it away.

- **CLEAN UP AFTER YOURSELF.** Tidy up your lab area as soon as you finish the experiments.

# RECIPE 1:
# CLASSIC OOBLECK

Let's get this science party started with a bang! Here's a recipe for the greatest, most classic slime of all time. The stuff is called **oobleck**. It has been delighting kids (and adults, too) for decades. Now it's your turn!

## YOU NEED:

- 1 cup (240 ml) cornstarch
- Resealable plastic bag
- ½ cup (120 ml) water
- Plastic dropper

## HERE'S WHAT YOU DO:

1. Put the water and cornstarch into a plastic bag. Seal the bag.

2. Squeeze gently to mix the water and cornstarch together. This step may take several minutes, so be patient. The mixture is done when it feels almost hard, but still oozes.

3. Add a little more water if the mixture seems too dry. (To do so, you can use the plastic dropper that came with this kit.) Add a little more cornstarch if it seems too runny.

### A SLIME IS BORN

A children's author named Dr. Seuss invented the term "oobleck." Seuss's imaginary goop appeared in a 1949 book entitled *Bartholomew and the Oobleck.*

### OOBLECK FOR DINNER?

Cornstarch is a common cooking ingredient. It is used to thicken soups, gravy, and other liquid-based foods.

# EXPERIMENT 1:
# OOBLECK FUN

Okay, you've made your oobleck. Pretty cool, isn't it?
Now you can play with it – and learn a little science
at the same time!

## YOU NEED:

- Disposable cup
- One small plastic animal or other toy
  that you don't mind getting slimy
- Disposable spoon

## HERE'S WHAT YOU DO:

1. Pour the oobleck into a disposable cup. *Slowly* insert a disposable spoon
   into the oobleck, then *slowly* stir. No problem, right? Now stir a little
   faster . . . and a little faster . . . and . . . HEY! The spoon is STUCK!

2. Pull the spoon out slowly (yes, it'll work) and set it aside.
   Set a plastic animal on top of the oobleck and watch it sink,
   like a quicksand victim. When the toy is almost submerged,
   grab it and yank it upward as quickly as you can.
   HEY, STUCK AGAIN! What's going on here?

3. Here's one more test. Pour a blob of oobleck into the palm of
   your hand. Tilt your hand back and forth and watch the oobleck
   ooze around. Now close your fist suddenly. See and feel what
   happens to the slime.

## SLIME SCIENCE

Like all slimes, oobleck is a polymer, which means it has
long, stringy molecules. When you apply pressure to the
oobleck, those molecules lock together and refuse to
move. When you release the pressure, they start to flow
again. You're basically changing your slime from a liquid to
a solid and back again. It's like magic – SLIME magic!

# ODDBALL FLUIDS

Oobleck is undeniably amazing. It is not, however, unique. It is just one of a whole GROUP of peculiar substances called **non-Newtonian fluids**.

To understand non-Newtonian fluids, let's first discuss the fluids that are Newtonian. Way back in the late 1600s and early 1700s, a scientist named Isaac Newton was trying to define fluids. He noticed that most fluids act consistently. If temperature and pressure conditions stay the same, they always flow in exactly the same way. Water, apple juice, milk, and MOST other liquids happily obey this law.

But a few "rebel" fluids insist on doing things differently. In particular, they refuse to flow consistently. Under pressure, some of them – like ketchup and blood – flow faster than normal. Others – like oobleck – flow more slowly.

Scientists tried their best to categorize these fluids. They couldn't. So they took the easy way out. They gave slimes and other NONconforming fluids a NON-category: They're NON-Newtonian. In other words, they don't obey Newton's laws.

A LOT of slimes are non-Newtonian fluids. You've already seen (and played with) one. And guess what? You're about to discover two more fantastic non-Newtonian slime recipes. Make 'em . . . mess with 'em . . . LOVE 'em!

## Isaac Newton (1643-1727)

An English scientist named Isaac Newton tackled just about every branch of science under the sun. He studied physics, mathematics, astronomy, chemistry, and more. His work in these areas paved the way for future scientists and is still taught in schools today. A brilliant thinker, Newton is considered one of the most influential people of all time.

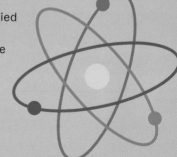

# A Non-Newtonian Lineup

Oobleck is a great example of a fluid that slows down under pressure. Many other slimes act the same way. Slime is NOT, however, the only non-Newtonian (oddball) substance! Here are a few other fluids that change their behavior in high-pressure situations.

## THESE FLUIDS SLOW DOWN (GET THICKER):

**Chilled caramel topping:** Eaten slowly, caramel is a gooey liquid. Poked HARD with a spoon, it's a solid. Hmmm, you'd probably better chew slowly as well. Your teeth will thank you!

**Silly Putty:** A ball of Silly Putty™ will flow into a puddle over time. Yank the ball apart quickly, though, and it will rip like a dinner roll. The molecules just can't move fast enough to keep up.

**Quicksand:** Quicksand acts a lot like oobleck. You can swim right out of a quicksand pit if you move slowly. If you struggle, though, the quicksand thickens. You'll be stuck forever!

## THESE FLUIDS SPEED UP (GET THINNER):

**Ketchup:** It can be tough to get ketchup moving. Once it starts, though, it flows easily. That's because the flowing process makes the ketchup thinner.

**Latex and oil paint:** Many paints are designed to thin out under pressure. This makes them easier to pick up and spread. When the pressure disappears, the paint thickens and (hopefully) doesn't drip all over the place.

**Human blood:** Under normal conditions, red blood cells clump together. But when pressure increases – inside tiny blood vessels, for instance – these clumps break apart. The blood becomes thinner and flows more easily as a result.

# MORE ODDBALL FLUIDS

Are you ready for a little more fun? You can make your own non-Newtonian (oddball) fluids with the recipes below!

## RECIPE 2: GLOOP

This recipe is super simple, but the results are incredibly cool. This slippery slime ball will entertain you and your friends for hours!

### YOU NEED:

- ½ cup (120 ml) white glue
- ¼ cup (60 ml) liquid starch (available in most grocery stores)
- Resealable plastic bag

### HERE'S WHAT YOU DO:

1. Put the white glue and the liquid starch into a plastic bag. Seal the bag.
2. Squeeze gently to mix the glue and starch together. You'll probably have to squeeze out a bunch of lumps. This step may take several minutes, so be patient.
3. Add a little more starch if the mixture seems too sticky. Add a little more glue if it does not flow. The mixture is done when you have created a big, smooth, solid ball of slime.

### SLIME SCIENCE

White glue is a liquid polymer, which means (of course) that it has all those long molecules. Liquid starch clings to those glue molecules. It's like adding a bunch of melted cheese to a spaghetti pile. **Picture that for a moment.** The gooey cheese holds everything together so there's no WAY those pasta pieces can wiggle around. Your slime ball? Just like that.

# RECIPE 3:
# MINTY FRESH

You just made a putty-like slime. Now let's try a really sticky one! This non-Newtonian goo will cling to your fingers like no one's business. But it smells GREAT, thanks to the addition of some minty-fresh toothpaste.

## YOU NEED:

- 2 teaspoons (10 ml) white toothpaste
- 4 teaspoons (20 ml) white glue
- 8 teaspoons (40 ml) cornstarch
- 2 teaspoons (10 ml) water
- Disposable cup
- Disposable spoon
- Plastic dropper

## HERE'S WHAT YOU DO:

1. Put the toothpaste, the glue, and the cornstarch into a disposable cup. Use a spoon to mix the ingredients together.

2. Add the water. Stir again to mix.

3. Use the dropper to add a few more drops of water if the mixture seems too dry. Add a little more cornstarch if it seems too runny. The mixture is done when it is smooth and gooey.

### SLIME SCIENCE

All those long molecules in this concoction are clinging together like crazy. They can still slide around, but the finished slime is SUPER sticky. That's because the molecules grab onto tiny imperfections in the surfaces they touch. It's like a million little anchors working together to keep the slime in place.

# SURROUNDED BY SLIME

At this point, you have a pretty good understanding of slime structure and behavior. But why stop now? It's time to get your hands on something *really* special.

How special? Well, the slimes that come with this kit contain a few surprises. These slimes are actually slime bases with amazing added ingredients. These ingredients make your slime wiggle, change color, and even glow in the dark. It's super slimy AND super cool!

These effects may be incredible, but the science behind them is very simple. Remember those polymer strands and the way they tangle together? It might not surprise you to learn that these strands can trap blobs of other substances. Chemically speaking, the blobs aren't part of the slime. They're just stuck, like meatballs in a pasta pile.

The next experiment takes advantage of this fact. It uses the properties of the slime *and* the extra-super-special ingredient. Try it and see!

## LINGO TIME

Substances that trap and support undissolved particles are called **suspensions**.

12

# EXPERIMENT 2:
## IT'S MAGNETIC!

Can slime be magnetic? Find out in this easy but incredibly cool experiment.

### YOU NEED:

- Packet of gray slime from this kit
- Several magnets of varying strengths
- Magnifying glass

### HERE'S WHAT YOU DO:

1. Lay the slime packet on a flat surface. *Do not open the packet*.

2. Use a magnifying glass to examine the slime. Can you see that the slime contains many tiny particles?

3. Set a magnet on the slime packet and leave it there for about a minute. Then remove the magnet.

4. Examine the slime again. Do the particles look different now?

### SLIME SCIENCE

The slime is gray because it holds lots and lots of tiny metal particles. These particles are magnetic. When you put a magnet on the slime, the particles want to line themselves up with the "pull" they feel. The slime molecules slowly rearrange themselves to allow this shift. When all the movement is done, you can clearly see the particles' new alignment through a magnifying glass. The stronger the magnet, the stronger the effect!

# EXPERIMENT 3:
## SLIME IN MOTION

So are you just dying to get your hands on that slime? Here's your chance. We're going to open things up and get a little goopy!

## YOU NEED:

- Packet of gray slime from this kit
- Several magnets of varying strengths
- Scissors
- Petri dish from this kit

## HERE'S WHAT YOU DO:

1. Use scissors to carefully cut open one corner of the slime packet. Squeeze a small blob of slime into the bottom half of your Petri dish.

2. Hold a magnet against the underside of the Petri dish, right below the slime blob. Move the magnet around. Try moving it quickly, then slowly. What do you see?

3. Repeat step 2 with several magnets. Does anything different happen?

4. Set the Petri dish on its edge by carefully leaning the dish against a wall or another upright object, making sure it won't roll away. Leave it alone for at least twenty minutes. Then look at the slime blob. Has anything changed?

### SLIME SCIENCE

This experiment uses two different forces to move your slime. In steps 2 and 3, a force called **magnetism** acts on the slime's metal particles. The particles move, and they drag all those polymer molecules along for the ride. The stronger the magnet, the stronger the effect. In step 4, a force called **gravity** comes into play. Gravity pulls the slime and the particles equally. The result is a slow downward ooze that's guaranteed to delight any slime fan.

14

# EXPERIMENT 4:
## IT'S ALIVE!

Did you think the last experiment was awesome? Well, you haven't seen *anything* yet. This one will knock your socks off!

### YOU NEED:

- Packet of gray slime from this kit
- Several magnets of varying strengths
- Petri dish from this kit

### HERE'S WHAT YOU DO:

1. Squeeze a little more slime into the Petri dish. (For best results, you want a fairly good-size blob.) Set the Petri dish on a flat surface.

2. Hold a magnet above the Petri dish. *Slowly* move the magnet downward, toward the slime blob.

3. At some point you will see the slime blob twitching. If you move the magnet just a hair closer, you should see a finger of slime erupt upward!

4. Now comes the *really* fun part. Carefully move the magnet up and down, back and forth. The slime finger will respond to the magnet's motion. Make that baby dance!

5. Repeat steps 2, 3, and 4 with several magnets of varying strengths. What do you see?

### SLIME SCIENCE

You already know that magnets attract the metal particles within the slime. When you hold a magnet above the slime, the particles want to LEAP upward. They do, to a point – but the slime holds them back. Those long polymer strands oppose the magnet's pull. By tweaking the distance of the magnet, you can find the perfect balance between these forces. The metal particles stay up, and the slime doesn't break. **Slime finger**, anyone?

15

# RECIPE 4:
# THICKENING SLIME

So you've seen one of your special slimes in action. Are you ready to see what lurks in packet number two? Of course you are! You have to make this one, so it's time to head back to the lab!

## YOU NEED:

- Test tube from this kit
- Disposable cup
- Wooden stirring stick from this kit
- Packet of color-changing powder (white powder)
- Scissors
- Paper towel
- Water

## HERE'S WHAT YOU DO:

1. Pour two full test tubes of water into a disposable cup.

2. Carefully cut open the packet of white powder. Dump all of the powder into the cup.

3. Use the wooden stick to stir the water and powder for about five minutes. Yep, FIVE WHOLE MINUTES. Time yourself.

4. Remove the wooden stick. Use a paper towel to wipe it off.

5. Let the slime sit for another ten to fifteen minutes. The longer it sits, the thicker it will get. Ewwww!

### SLIME SCIENCE

It takes plenty of time and contact for the molecules in this polymer mixture to bond. That's why you have to stir, and stir, and stir, and then wait. Once everything finally starts to join up, however, there's no turning back. Your goo gets thicker and thicker as more and more molecules connect.

# EXPERIMENT 5:
## WHY THE CHANGE?

The slime you just made is plain white. It might seem – dare we say it? – kind of BORING. But that's about to change in a really colorful way. Let's find out what flips your slime's "on" switch!

### YOU NEED:

- Wooden stirring stick
- Different lighting types and conditions
- White foam or plastic plate
- White slime

### HERE'S WHAT YOU DO:

1. Use the wooden stick to scoop a dime-size blob of slime onto the plate. (Make sure the plate is white. A white background will make it easier to see what happens next.)

2. Take the plate outdoors. Expose the slime to direct sunlight for about one minute. Do you see any difference? *(Hint: You should!)*

3. Expose the slime to other types of light. Try weak or indirect sunlight, regular light bulbs, fluorescent light bulbs, your computer screen, a TV screen, an LED light, and anything else you can find. See what effects, if any, these light sources have on your slime.

### SLIME SCIENCE

The slime you made on page sixteen contains dye molecules. These molecules are activated by ultraviolet light, which is also called UV. Sunlight contains a great deal of UV. Direct sunlight, therefore, has the strongest effect on your slime's color. Weak sunlight carries less UV and has less effect. Indoor lighting carries very little UV and causes little to no color change. What about the other types of light we mentioned in step 3? Try them and see for yourself!

# EXPERIMENT 6: JUST PASSING THROUGH

Now you know how to make your slime change color. In this experiment, your goal is exactly the opposite. You want to *stop* it from changing. Let's see if we can accomplish this sensational sun-blocking feat!

## YOU NEED:

- Two white foam or plastic plates
- Petri dish and lid from this kit
- The color-changing slime made in experiment #4
- Clear, spray-type sunblock, SPF 30 or higher
- Wooden stirring stick

## HERE'S WHAT YOU DO:

1. Use the wooden stick to scoop one fresh dime-size blob of slime onto each plate.

2. Spray one blob with sunblock. Leave the other one alone.

3. Scoop another blob into your Petri dish. Put the lid on the dish.

4. Carry both plates and the Petri dish outside. Let them sit in direct sunlight for about one minute. What do you see?

### SLIME SCIENCE

The uncovered, unsprayed blob should change color right away. The blob in the Petri dish should change a little bit, but not as much. And the sunblock-sprayed blob shouldn't change at all. What's going on here? Well, the sunblock protects the slime from the sun's UV rays, just like it protects your skin from sunburn. The Petri dish blocks *some* UV, but not all of it. Your slime tells the tale!

# COLOR-CHANGING MAGIC

You already know that your color-changing slime contains dye molecules. But how exactly do those molecules create the color shift you see? It's a pretty cool trick . . . and LUCKY YOU, we're going to explain it right now!

Here's the scoop. Yes, your slime contains dye molecules. Those dye molecules, however, are extremely tiny under normal conditions. They're SO small, in fact, that they can't trap visible light (which is different from UV). This means the dye molecules are colorless.

But things change when UV rays show up for the party. The UV rays add energy to the dye molecules, which start to cling together. Soon the dye clumps get big enough to trap visible light. The trapped light can be seen as a color.

This change isn't permanent. When the UV goes away, the molecules get all lazy again. They fall apart, and the color disappears. They'll be more than happy to do the whole performance all over again, though. Just find a little sunlight and OFF YOU GO!

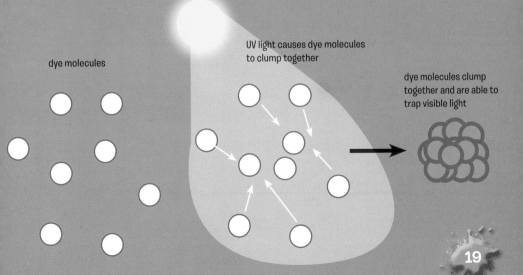

dye molecules

UV light causes dye molecules to clump together

dye molecules clump together and are able to trap visible light

# RECIPE 5:
# GLOW-IN-THE-DARK DIAPER SLIME

Are you ready for yet another cool suspension? In *this* slime, the suspended particles actually glow in the dark. Spooky!

## YOU NEED:

- Test tube from this kit
- Large resealable plastic bag
- Packet of green powder from this kit
- Wooden stirring stick from this kit
- Clean, unused diaper
- Disposable cup
- Scissors
- Water

## HERE'S WHAT YOU DO:

1. Use scissors to cut a diaper into small chunks. Put the chunks into a large resealable plastic bag.

2. Seal the bag. Shake it for several minutes.

3. Open the bag and remove the diaper pieces so that only the diaper powder is left. Throw the pieces away.

4. Carefully cut open the packet of green powder from this kit. Pour green powder into the test tube until it reaches the bottom line.

5. Pour the green powder from the test tube into a disposable cup. Add one full test tube of water and one good-size pinch of diaper powder. (Save the leftover powder. You'll need it in another activity.)

6. Use the wooden stick to stir everything thoroughly. After about 30 seconds, you'll have a cupful of disgusting green SLIME! You

# EXPERIMENT 7:
# TAKE-ALONG GLOWING SLIMESTICK

Yes, your diaper slime really DOES glow. You can use this power for nighttime visibility. Here's how to do it!

## YOU NEED:

- Test tube and cap from this kit
- Glow-in-the-dark diaper slime from page 20
- Wooden stirring stick

## HERE'S WHAT YOU DO:

1. Use the wooden stick to scoop glow-in-the-dark diaper slime into the test tube. Fill the tube to the brim, then screw the lid on.

2. Leave the test tube in a brightly-lit place for at least five minutes. Direct sunlight works best, but any bright light will do.

3. Carry the test tube into a dark place, such as a closet or a bathroom without any windows – the darker, the better. Marvel at the eerie green glow!

4. After a while, the glow will fade. Just expose the test tube to light again to get things glowing again. The fun never ends with this rechargeable slime!

### SLIME SCIENCE

The glow process in this experiment is pretty simple - simply AWESOME, that is! The green powder contains molecules that absorb light energy. They do this constantly when light is available. When it's dark, though, the molecules start to release their stored energy. The released energy can be seen as visible light. This type of glow is called phosphorescence.

21

# EXPERIMENT 8:
## SALTY SLIME

Diaper slime is pretty sturdy. You can stir it, shake it, or even sit on it without hurting it a bit. One common substance, however, can MELT YOUR SLIME AWAY. You're about to see this process in action!

### YOU NEED:

- ¼ teaspoon (1.25 ml) of diaper powder from page 20
- ½ cup (120 ml) of water
- Wooden stirring stick
- Salt
- Disposable cup
- Plastic or foam plate

### HERE'S WHAT YOU DO:

1. Put the diaper powder into a disposable cup. Add the water and stir with the wooden stick to mix. Let the mixture sit for about five minutes to reach maximum sliminess.

2. Use the wooden stick to scoop out several big lumps of slime. Pile them on a plastic plate.

3. Shake salt all over the slime. Really cover it! Watch and see what happens.

### SLIME SCIENCE

The salt breaks many of the bonds between the diaper powder's polymer molecules and its water molecules. The slime gets wetter and wetter as more and more bonds break and more water escapes. After a while, the slime may even start to OOZE water around the edges. Come to think of it, what's slime without a little ooze?

# CREATURES OF SLIME

Did you know that diaper slime has something in common with SLUGS? It's true! Just like diaper slime, slugs cannot tolerate salt. It damages their sensitive sluggy skin.

Want some more slimy critter trivia? We've got you covered! Here's the 411 on nature's EWWW-iest, gooiest citizens.

- **Snails** glide on a cushion of slime! They constantly ooze mucus from their bellies. The mucus coats uneven surfaces and prevents the snail from hurting its squishy self.

- **Hagfish** use slime to defend themselves! These eel-like fish spew a substance that expands rapidly into BUCKETS of slime when it hits the water. This slimy material traps, blocks or, sometimes, entangles attacking creatures.

- **Green moray eels** are actually blue! They look green because they're coated with a layer of thick yellow slime. Yellow plus blue equals GREEN, so that's the color you see!

- **Worms** ooze slime from their bodies. The slime contains a chemical that fertilizes plants. Worms as plant food? Slimy but SUPER!

- **Velvet worms** (which are different from earthworms) are carnivores that SHOOT their prey with slime! They squirt the entangling goo out of holes on both sides of their mouths to catch small insects like termites, crickets and spiders.

# A REALLY DENSE SUBJECT: VISCOSITY

Now that you've made a whole BUNCH of different slimes, we're ready to tackle a very big word. That word is **viscosity,** and it's an essential concept in the slime world.

So what exactly is viscosity? Here's the hardcore scientific explanation: It's a measurement that describes how hard the molecules of a fluid rub against one another. Slimes (and other fluids) with a low "rub" factor have low viscosity. Oobleck, for example, has a fairly low viscosity compared to most of the slimes in this book. Slimes with a higher "rub" factor have higher viscosity. The gloop from this kit, for example, has a fairly high viscosity.

Measuring viscosity – *really* measuring it – requires all kinds of fancy scientific equipment. Luckily, you don't have to worry about this. You can simply apply a very UNscientific rule of thumb: the thicker the goop, the higher the viscosity.

A substance's **flow rate** provides a great demonstration of this principle. In general, highly viscous (that is, thicker) stuff flows more slowly than less viscous (that is, thinner) goo. Want to prove it? YOU CAN. Let's do an easy experiment to show viscosity in action!

## POP QUIZ!

Scientists have measured the viscosity of these flowing foods. Can you guess the results? Put these substances in order from LEAST viscous (runniest) to MOST viscous (thickest).

\_\_\_\_ Ketchup

\_\_\_\_ Corn syrup

\_\_\_\_ Molasses

\_\_\_\_ Chocolate syrup

\_\_\_\_ Olive oil

\_\_\_\_ Peanut butter

\_\_\_\_ Honey

\_\_\_\_ Water

Answers:
1. Water; 2. Olive oil; 3. Corn syrup; 4. Honey; 5. Molasses; 6. Chocolate syrup; 7. Ketchup; 8. Peanut butter

# EXPERIMENT 9:
## MOVING AT THE SPEED OF SLIME

It's time for slime – a slime RACE, that is! Start your scientific engines, because it's time to see which slime will win the Viscosity Challenge.

### YOU NEED:

- Any (or ALL) of the slimes from this kit
- Any timing device (a watch, a clock, a sand timer, etc.)
- Cookie sheet

### HERE'S WHAT YOU DO:

1. Set a cookie sheet on a flat surface. Squeeze, pour, or scoop a nickel-size blob of each slime onto the sheet. The blobs should form a line across one edge of the cookie sheet.

2. How fast do you think each blob will flow? Record your guesses in the chart on page 27.

3. Lift the blob-lined edge of the cookie sheet until the tray tilts sharply. Prop the tray up against a wall or another sturdy object.

4. Start your timer! Observe your slime lineup for at least fifteen minutes. See which slimes race to the bottom of the cookie sheet and which ones stall at the starting line. Record your results on page 27.

### SLIME SCIENCE

Gravity pulls on the slime blobs when you tilt the cookie sheet. The blobs start to flow downward – but they don't all flow at the same rate. Blobs with low viscosity flow quickly. Blobs with high viscosity flow more slowly.
By comparing the blobs' travel rates, you can easily rank their viscosity. No fancy measurements required!

# EXPERIMENT 10:
## GOOP DROP

This experiment tests your slime's viscosity in a totally different way. Get ready to drop the ball – literally!

### YOU NEED:

- Any (or ALL) of the slimes from this kit
- One small disposable cup (bathroom-size or smaller) for each slime
- One identical marble or metal ball for each slime
- Any timing device (a watch, a clock, a sand timer, etc.)

### HERE'S WHAT YOU DO:

1. Squeeze, pour, or scoop a generous amount of slime into each small cup. (Hint: You might be running low on the slimes that came with this kit. THAT'S OKAY. Just use whichever ones you have. You could also fill a few cups with household substances, like ketchup or chocolate syrup, if you want to expand your experiment.)

2. How fast do you think marbles will sink into each substance? Record your guesses in the chart on page 27.

3. Start your timer, or check a clock and make note of the exact time.

4. Drop a marble or ball into each cup. Work quickly!

5. Now sit back and watch what happens! Which slimes suck the marbles right down? Which ones barely budge? Record your results on page 27.

### SLIME SCIENCE

You know the drill: It's all about viscosity. Thinner slimes can rearrange their molecules quickly. They allow marbles to sink right in. Thicker slimes take a while to get out of the way. Marbles sink more slowly in these materials.

## CHART: Moving at the Speed of Slime

Write down the names of the slimes you're using for Experiment 9. Fill in their expected and actual ranks. Use the "1" for the fastest slime, "2" for the second-fastest slime, and so on.

| Slime name: | Expected rank: | Actual rank: |
| --- | --- | --- |
| _____ | _____ | _____ |
| _____ | _____ | _____ |
| _____ | _____ | _____ |
| _____ | _____ | _____ |
| _____ | _____ | _____ |

## CHART: Goop Drop

Write down the names of the slimes and household substances you're using for Experiment 10. Fill in the expected and actual drop time of marbles in each material.

| Slime name: | Expected rank: | Actual rank: |
| --- | --- | --- |
| _____ | _____ | _____ |
| _____ | _____ | _____ |
| _____ | _____ | _____ |
| _____ | _____ | _____ |
| _____ | _____ | _____ |

So how did you do? Did you guess correctly? If so, GOOD JOB. If not, that's okay! You learned something, which means you're a winner no matter what!

# JUST FOR FUN:
## SLIME ART

You've been working so hard. You deserve a little fun – and you're about to get it! Let's make a couple of slimes that you can use in your next art project.

## RECIPE 6: SLIME GLUE

This simple glue is just as fun to make as it is to use!

### YOU NEED:

- White vinegar
- 1 tablespoon (15 ml) of fat-free milk
- Plastic or foam plate
- Disposable cup
- Paper towel

### HERE'S WHAT YOU DO:

1. Pour vinegar into a disposable cup until it just cover the bottom.

2. Add the fat-free milk. Swirl the cup to mix the liquids. You will see the milk breaking into stringy chunks.

3. Put a paper towel on a plate. Pour the milk/vinegar mixture onto the towel. The paper towel will soak up the vinegar, leaving a slimy white substance behind.

### SLIME SCIENCE

Vinegar has an immediate effect on a milk protein called **casein**. It makes the casein **polymerize**, which means it forms long molecules. Those molecules cling together and turn all of the milk's casein into a sticky blob. This natural "plastic" is one of nature's best glues. Try it and see for yourself!

28

# RECIPE 7: SLIME PAINT

How about a little paint to go with your glue?
Let's get artistic!

## YOU NEED:

- 2 tablespoons (30 ml) liquid starch
- 2 tablespoons (30 ml) clear shampoo
- 2 tablespoons (30 ml) flour
- Food coloring
- Disposable cup
- Disposable spoon

## HERE'S WHAT YOU DO:

1. Put the liquid starch, the shampoo, and a few drops of a food coloring into a disposable cup.

2. Use the spoon to mix everything together.

3. Add the flour and stir again. Use the spoon to smash the flour blobs apart, if necessary. Your mixture needs to be creamy-smooth. You made paint!

**Want different color paints?**
**Repeat this activity using another food coloring.**

## SLIME SCIENCE

This polymer has a very low viscosity compared to some of the others you've made in this kit. For this reason, it is easy to pour and spread. It dries with an interesting gritty texture that you'll just love!

## A CHEMICAL CHANGE

Want to see something totally different? **WITH A PARENT'S HELP AND PERMISSION**, microwave a batch of slime paint for thirty seconds. (Make sure you use a microwave-safe container.) We don't want to give away the surprise, but let's just say you'll see a big change. The heating process causes a chemical reaction that dramatically alters your slime!
**Do NOT use the magnetic (gray) slime for this activity.**

# STUFF TO TRY

This book is almost over. But the slimy fun doesn't have to end! It's time to put on your "scientist" hat and figure out some more cool stuff to try with your ooey, gooey concoctions. Here are a few ideas to get you started.

**Change your slime's temperature.** Chill it in the fridge, freeze it in the freezer, or heat it under the hot sun. See what (if anything) happens to the slime as these changes occur.

**Create a mold garden.** Can your slime grow mold? Test it and see! Contaminate a blob of slime with dirt. Seal the slime in a plastic bag for a few days and see if anything grows.

**Dry the slime.** Does your slime dry out if you let it sit in the open air? Try it and see.

**Wet the slime.** What happens if you add water to your slime? Does it absorb the water at all? If so, does the water improve the slime's texture? Or does it just create a runny mess?

**Mix the slimes.** Try combining small blobs of different slimes. Observe what happens. Try to figure out the science behind the results.

**"Copy" the newspaper.** Some slimes (generally the putty-like ones) pick up newsprint. Find out if any of your slimes do this cool trick. Press the slime against a sheet of newspaper. Gently peel it off and see if you've captured anything newsworthy!

# SLIME HUNT

The BEST slime activity of all, of course, is to find MORE SLIME! With a parent's permission and help, hunt for gooey stuff in your kitchen cabinet, your refrigerator, your bathroom, your backyard, and anywhere else you can imagine. You might find slimy foods, growths, or lotions. You might even discover some long-forgotten, liquid leftovers. GROSS!

Once you've found some slime, it's science time! Do you think your slimes are polymers? Are they alive (like bacteria, for example)? Are they Newtonian or non-Newtonian? Use the Internet, the library, your science teacher, or any other reference source to find more information.

Hey, this sounds like it could be a SOCIAL OCCASION! Invite a friend over for a slime hunt. See who can find (and identify) the coolest stuff. Anyone could win – but in this contest, you have a BIG head start. You've learned a lot about polymers, and non-Newtonian fluids, and viscosity, and so many other slime-related topics. Put your knowledge to use. With a little creativity, you're sure to be the ULTIMATE slime scientist!

# GLOSSARY

**flow rate**          The volume of liquid (or fluid) which passes through a specified surface in a given amount of time.

**molecule**           A tiny unit of matter made from two or more linked atoms.

**non-Newtonian fluid**   A fluid that does not act according to Newton's laws.

**oobleck**            A basic slime created by mixing water and cornstarch.

**phosphorescence**    Light created in dark conditions by the release of stored light energy.

**polymer**            A substance with long, chainlike molecules.

**polymerize**         To become a polymer.

**suspension**         Substances that trap and support undissolved particles.

**ultraviolet (UV) light**   A high-energy light. Sunlight contains UV, which is invisible to the human eye.

**viscosity**          A fluid's resistance to flow. Greater resistance equals greater viscosity. Lower resistance equals lower viscosity.